The Question & Answer Book

OUR WONDERFUL SOLAR SYSTEM

OUR WONDERFUL SOLAR SYSTEM

By Richard Adams
Illustrated by Ray Burns

Troll Associates

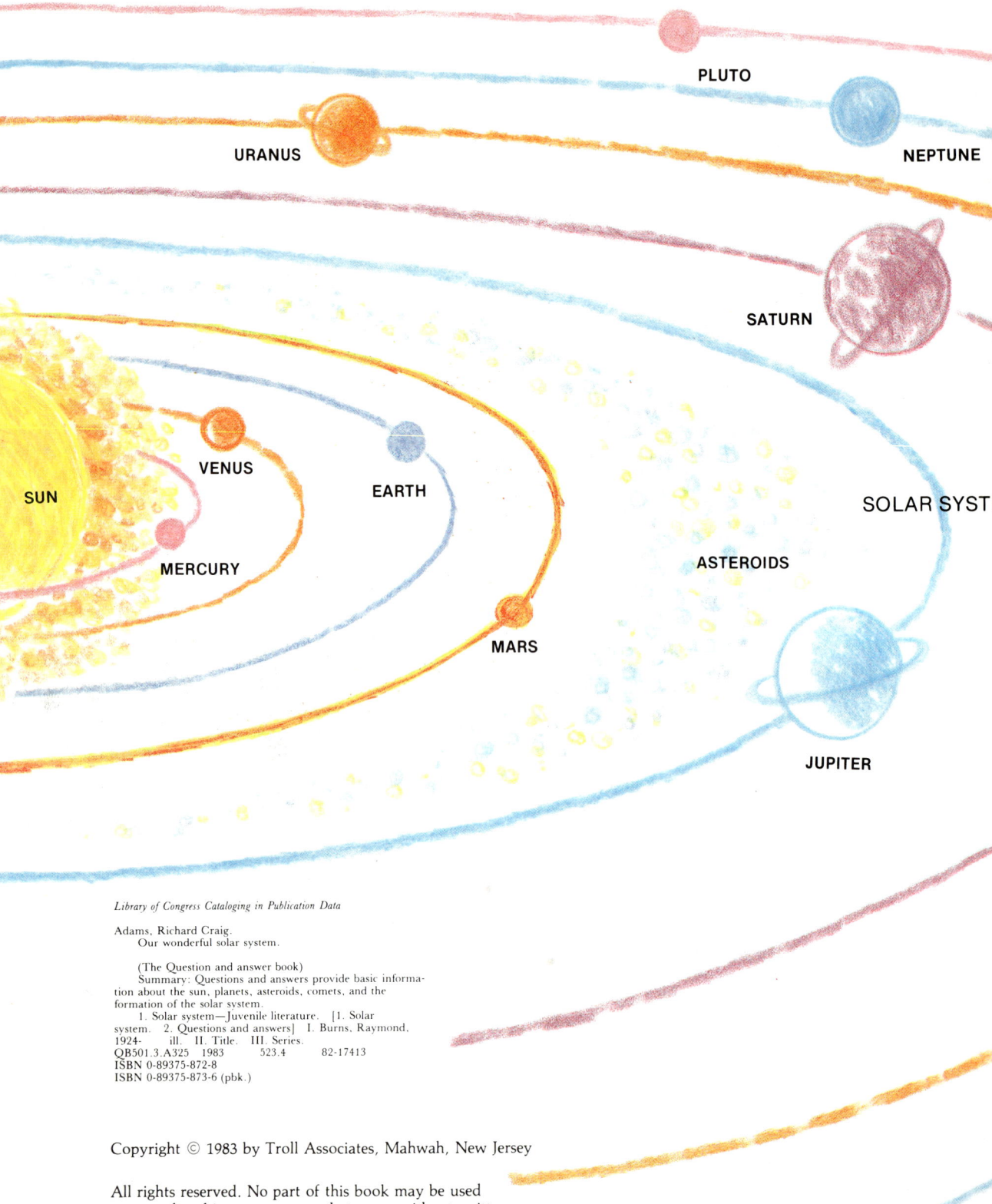

Library of Congress Cataloging in Publication Data

Adams, Richard Craig.
  Our wonderful solar system.

  (The Question and answer book)
  Summary: Questions and answers provide basic information about the sun, planets, asteroids, comets, and the formation of the solar system.
  1. Solar system—Juvenile literature. [1. Solar system. 2. Questions and answers] I. Burns, Raymond, 1924-   ill. II. Title. III. Series.
QB501.3.A325 1983   523.4   82-17413
ISBN 0-89375-872-8
ISBN 0-89375-873-6 (pbk.)

Copyright © 1983 by Troll Associates, Mahwah, New Jersey

All rights reserved. No part of this book may be used or reproduced in any manner whatsoever without written permission from the publisher.

Printed in the United States of America
10 9 8 7 6 5 4 3 2

Can you imagine?

Imagine you are far, far out in space in a rocket ship. All is dark around you except for the brightly shining stars. When you look about, you can see one special point of brilliant light—the sun. Gathered around the sun are nine round, shining objects. What could they be? They are the planets. Some seem to be near the sun. Others are very far away from it. If you look hard, you can see that the third planet from the sun is your home, the Earth. The other planets are Mercury, Venus, Mars, Jupiter, Saturn, Uranus, Neptune, and Pluto.

You are looking at what scientists call the *solar system*. It is made up of the sun, the planets, and many smaller objects.

The sun is the biggest object in our solar system.

The sun is also the most important. You might say the sun is the "star" of the solar system. And you would be right in more ways than one. For the sun really *is* a star! It is a star just like all those you see in the sky at night. From Earth, the sun looks bigger than the other stars because it is much closer to us.

Actually, the sun is only a middle-sized star. Some stars are much, much bigger. Other stars are smaller. But they are all made up of very hot, glowing gases. You could not get very close to any of them. Stars also give off a great amount of light as well as heat.

But, except for Jupiter, the planets do not produce their own heat or light. They receive all their light and heat from the sun.

Then why do the planets seem to be shining? The light that the planets give off is a reflection of the sun's light. That means the sun's light rays hit each planet and bounce off it again.

What if there were no sun?

Imagine what the Earth would be like if there were no sun. There would be no daytime—only the darkness of night. And the Earth would be very, very cold. It would be much too cold to live on.

If you could watch the planets from your rocket ship in outer space, you would see that they are all constantly moving around the sun. Each planet moves in a special path called an *orbit*. Each orbit is oval-shaped—it looks something like the outline of an egg. Every planet stays in its own orbit—so it does not crash into another planet.

How long does it take the Earth to make one orbit?

The Earth takes one year—365 days—to make one complete orbit around the sun. The other planets take more time or less time. The planet Pluto takes nearly 248 of our Earth-years to orbit the sun.

Pluto's trip around the sun takes a long time because Pluto is very far away from the sun. Pluto must travel a tremendous distance to make one orbit. The planet Mercury takes less than three months to make a full orbit. That is because Mercury is very close to the sun—the distance Mercury must travel is far less than the distance Pluto must cover.

Mercury

Mercury is the planet closest to the sun. Its days are very hot. Imagine the oven in your kitchen turned up as high as it can go. The temperature of Mercury during the day is much hotter. But at night, the temperature is very, very cold.

Mercury is not only hot, it is very dry. There is no rain on Mercury. And there is little air there. Mercury's surface looks much like our moon. It seems to be a rocky world. Its surface is covered with deep holes called *craters*. We know this because a spacecraft flew very close to Mercury and took pictures. This spacecraft also measured the planet. Now we know that Mercury measures 3,100 miles (4,960 kilometers) straight through its middle. That's about the same as the distance between New York City and San Francisco, California.

For many years, scientists believed that Mercury was the smallest of the nine planets. But now they are not so sure. Some think Pluto may be smaller. But Pluto is very far away from Earth. Scientists cannot see it clearly, even when they look through the most powerful telescopes. In the future, space probes may help us find out more and more about all the planets.

Venus

    Venus is the second closest planet to the sun. It is very hot—even hotter than Mercury. It averages about 870° Fahrenheit (465° Celsius). Because of the extreme heat on Venus, no oceans now exist on that planet. Any water would be turned to steam immediately. Did you ever see someone sprinkle water on a hot iron? With a sizzle, the water disappears instantly. It becomes steam and goes up into the air. That's just what would happen to water on Venus.

    Venus has thick clouds all around it. These clouds probably act as a blanket, keeping the sun's heat trapped under them. That is why the temperature on Venus stays high all the time.

By sending spacecraft equipped with radar to orbit Venus, scientists are beginning to gather information about this planet's surface.

Venus is sometimes called Earth's sister planet. This is because the two planets are alike in several ways. Venus comes closest to Earth in its orbit. It is nearly the same size as Earth. And its air is made up of some of the same gases as ours.

Could life exist on Venus?

Scientists do not think so—at least, not life as we know it. None of the animals that live on Earth could live on Venus. The air on Venus has no oxygen. Like people, animals need to breathe oxygen. The air on Venus *does* have the gas carbon dioxide, which plants need to live. But none of the plants that live on Earth could live on Venus. It is much too hot.

Earth

The next planet from the sun is our own Earth. What makes Earth special is that life can exist on it. Earth has oxygen for people and animals to breathe. It has carbon dioxide for plants to breathe. It has water for plants and animals to drink. And its temperature, in most places, is not too hot or too cold for life.

Moving around and around the Earth is the moon. It goes around the Earth in an orbit—just as the Earth goes around the sun in an orbit. The moon makes one complete trip every month.

We know more about the moon than we do about any of the planets except Earth. That is because the moon is closer to Earth than any of the planets. So we have been able to send people to the moon. Astronauts have walked on the moon and brought back moon rocks for scientists to study.

Now we know that the moon is a rocky place full of high mountains and deep craters. There are millions of craters on the moon. The smallest are less than one foot (30 centimeters) wide. Some are five miles (8 kilometers) wide. Others are much larger. One of these large craters is about 700 miles (1,100 kilometers) wide!

On the moon, there is no air to breathe. The moon has no atmosphere. So astronauts have to carry tanks of oxygen with them. They also find it easier to hop than to walk on the moon. That is because the moon has little *gravity*, or downward pull.

15

Like the planets, the moon doesn't give off any heat or light. Then why do we see the moon shining at night? It is reflecting the light of the sun, and so it seems to shine.

Did you know that there is more than one moon in the solar system?

We call any object that circles a planet a moon. And the Earth is not the only planet to have an object orbiting it. Scientists have found that other planets have moons, too—from one to more than twenty.

Mars

Earth's second closest neighbor in space is Mars. Although it is smaller than Earth, Mars is more like Earth than any other planet. The temperature on Mars is colder than on Earth. But many scientists believe that Mars is warm enough to support life. Mars has a tiny bit of oxygen in its air. It has a lot of carbon dioxide in its air, too. Many scientists think that Mars probably once had water. And there may be water there now—frozen both upon and beneath its surface. If this is so, life—probably plant life—could exist on Mars.

Unmanned spacecraft have landed on Mars. They tested its soil with delicate instruments. No living things were found in the areas tested.

Photos taken of Mars show that its surface looks something like our moon. It is rocky and has craters. Unlike the moon, Mars has the remains of old volcanoes on its surface. It also has other marks that our moon does not have. These are channels that look like dried-up river beds. Many scientists believe that if there were ever rivers on Mars, that planet once had a lot of water. This means that life may have existed on Mars at that time.

Besides its channels, Mars also has deep valleys called *canyons*. They may also have been cut by flowing water. One Martian canyon is ten times longer and three times deeper than the Grand Canyon on Earth!

Mars has a North and South Pole—just as Earth does. And as on Earth, the poles are very cold places. At each pole is a white area called a *polar cap*. Scientists think these caps are made of ice. During a Martian summer, the polar caps get smaller—as if they have melted. During the winter, they get bigger—as if they have frozen again.

Mars has two moons. But these moons are not round, like Earth's moon.

Jupiter

Jupiter, Mars' next neighbor, has at least sixteen moons. Each of these moons travels in a path around Jupiter. It's amazing that they never bump into each other! Some of Jupiter's moons are very small. Its largest moon is about 3,160 miles (5,270 kilometers) wide. Jupiter's largest moon is bigger than the planet Mercury. Scientists have seen volcanoes erupting on Io, one of Jupiter's moons.

Jupiter is the biggest planet in the solar system. Straight through its center, it is more than eleven times as wide as Earth.

Like Venus, Jupiter is covered with thick layers of clouds. In these clouds is the famous Great Red Spot. This spot is probably an area of stormy gases that swirl around and around. The spot is so big that the entire planet Earth could fit inside it. In 1979, a spacecraft passed Jupiter, giving scientists the first evidence of the narrow ring that surrounds this planet.

Most scientists believe Jupiter has a core of hot liquid that is about the size of Earth. This core is hidden far beneath layers of swirling gases.

Jupiter sounds like a terrible place for people to visit, and not only because of the hot liquid surface. Above Jupiter's clouds, it is very, very cold—about 184°F below zero (120°C below zero). The air thermometer in your house or classroom does not even go that low. Nowhere on Earth does the air ever get that cold. And far below Jupiter's clouds, the temperature is much hotter than anywhere on Earth.

Jupiter's heat comes more from its own center than from the sun. In this way, Jupiter is different from all the other planets. It is more like a small star than a planet. The other planets get all their heat from the sun.

Saturn

The planet beyond Jupiter is another giant, Saturn. It is not quite as large as Jupiter. Saturn has many rings around it. The rings are made of millions of tiny chunks of rock and ice that travel around this planet. The rings are really millions of tiny moons. Besides its rings, scientists believe Saturn may have as many as twenty-three full-sized moons.

Saturn is a cold planet because it is so very far from the sun—almost twice as far as Jupiter. Saturn and the sun are more than 800 million miles (1 billion, 280 million kilometers) apart. At this great distance, Saturn takes nearly twenty-nine and a half Earth-years to make one trip around the sun.

Uranus, Neptune, and Pluto

The last three planets in our solar system are even farther from the sun than Saturn. So they are even colder than Saturn. The three are Uranus, Neptune, and Pluto. They are so far away from Earth that scientists don't know much about them yet. However, scientists *do* know that Uranus has five moons. Neptune has two. Pluto has one.

Every day, scientists are making exciting new discoveries about the solar system. For a long time, it was thought that only Saturn had rings. Then scientists made the important discovery that Jupiter and Uranus also have rings! Scientists believe that Neptune—and perhaps Pluto—may have rings, too. The rings are probably made mostly of ice because these planets are so far away from the sun's warm rays.

Because Pluto is so far from the sun, very little sunlight reaches it. Imagine yourself on Pluto looking at the sky. The faraway sun would look as small as any other star.

Pluto may be the smallest major planet in our solar system. But because it is so far away, no one is sure of its exact size.

Because Neptune and Pluto are very far from Earth, you cannot see them without a telescope. But it is possible to see all the other planets with the naked eye. Uranus and Mercury, however, are very hard to see. Uranus is hard to see because it is so far away. Mercury is hard to see because it is so small and close to the brilliant sun.

Planet or star?

You can usually tell if you are looking at a planet or a star in the sky. A planet looks like a star, but it does not twinkle like a star. Its light is steady. Also, a planet changes its place among the stars from night to night. The stars keep their places beside each other. Just as a circle of children who are all moving together keep their places, so do the stars. But a planet is like a child who is *not* part of the circle, moving around it alone.

The easiest way to find a planet in the sky is to look for the "evening star." This is the first bright steady light in the sky in the evening. In spite of its name, the evening star is not a star at all. It is a planet. Which planet it is depends upon the time of year.

What are asteroids?

Besides the nine major planets, there are thousands of other, much tinier planets in our solar system. These are small chunks of rock called *asteroids*. Some people call asteroids "mini-planets." The tiniest asteroids are less than a mile (1.6 kilometers) wide. The largest one known is nearly 500 miles (800 kilometers) wide. Most asteroids orbit the sun in the area between Mars and Jupiter.

What are meteoroids?

Our solar system has some objects in it that are even smaller than asteroids. These are called *meteoroids.* Scientists think meteoroids fall through space near the Earth. But few land here. Most are burned up in the atmosphere during their trip down. As they fall, they look like streaks of light. People sometimes call them "shooting stars." The meteoroids that do land on the Earth are called *meteorites.* You can see them in a museum or planetarium. They look like chunks of rock. And except for their unusual shapes, most of them look just like the rocks on Earth.

Have you ever seen a comet?

A comet looks like a hazy star with a long tail following behind. Comets are another part of our solar system. Scientists believe they are probably made of frozen gases, ice, and bits of dust. Most of them have two parts—a head and a long tail. The head is a solid frozen ball—like a gigantic snowball. The tail is made of gases and dust that the sun's energy has pushed away from the head. The tail of a comet may be millions of miles long! But some comets have no tail at all. The others develop a tail only after they have passed near the sun.

Comets travel around the sun, just as the planets do. But the paths of comets are different from the paths of planets. The paths of the planets never cross each other, so they can never crash into each other. But the paths of comets sometimes cross the paths of planets. Then a crash *can* happen. In the year 1910, it is believed that the Earth and the tail of a comet called Halley's Comet met. People were afraid the world would come to an end. But the tail of Halley's Comet did not hurt the Earth at all. The Earth passed right through it unharmed.

Comets appear in our sky from time to time. They can be seen for a few weeks or a few months. Then they move out of our sight.

If you stand and watch a comet, it looks as if it isn't moving. But if you watch the sky for several nights, you will notice that the comet is in a different spot each night. So you know it has moved.

No one knows how many comets are in our solar system. Each comet takes a different number of years to make a full orbit. Halley's Comet takes about 77 years. That means people on Earth can see this comet once every 77 years. Comet Kohoutek, which appeared in 1973, takes 75,000 years to make a full orbit. Scientists know this by measuring the speed of the comet and the curve of its orbit. Then they are able to figure out how long it would take the comet to make a full trip around the sun.

How was the solar system formed?

Now you know that our solar system is made of the sun, nine major planets, many asteroids and meteoroids, and a number of comets. But where did they all come from? No one is sure of the answer. Many scientists believe that the solar system started as a mass of dust and gases in space. The mass was shaped something like a gigantic pancake. The flat mass started to spin. The center got very hot. It became the sun. Other pieces came together on the outside of this spinning mass. They became the nine planets and the larger asteroids. Other asteroids may have formed when two large ones crashed into each other. That is how meteoroids probably formed as well. The planets' moons may be pieces that broke off from the planets. No one knows how the comets formed. Many questions about the solar system are yet to be answered. Scientists are continually at work, trying to put together the many pieces of this immense puzzle.

Once again, imagine that you are in your rocket ship out in space. Now, as you look down at the sun and the objects around it, you can almost feel the heat coming from Venus—and the cold from Saturn. You can imagine Jupiter's hot liquid surface—and feel Mars' rocky landscape. Perhaps you can see the moons circling Uranus— and meteoroids shooting past. The solar system is a beautiful, mysterious world—and you're a part of it.